STEM Superstars

Neil deGrasse Tyson

by Rachel Castro

NorwoodHouse Press

Cover: Neil deGrasse Tyson teaches people about space.

Norwood House Press

For information regarding Norwood House Press, please visit our website at: www.norwoodhousepress.com or call 866-565-2900.

PHOTO CREDITS: Cover: © Charles Sykes/Invision/AP Images; © Arnold Gold/New Haven Register/AP Images, 12; © Craig Blankenhorn/Invision for FOX Broadcasting/AP Images, 18; © DenisTangneyJr/iStockphoto, 11; © Dennis Van Tine/Abacausa.com/Newscom, 21; © Desiree Navarro/WireImage/Getty Images, 15; © John Badman/The Telegraph/AP Images, 8; © Linda Cataffo/NY Daily News Archive/Getty Images, 17; © Pavel Gabzdyl/Shutterstock Images, 6; © TierneyMJ/Shutterstock Images, 5

Hardcover ISBN: 978-1-68450-834-1
Paperback ISBN: 978-1-68404-634-8

Library of Congress Cataloging-in-Publication Data
Names: Castro, Rachel, author.
Title: Neil deGrasse Tyson : astrophysicist / by Rachel Castro.
Description: Chicago : Norwood House Press, [2021] | Series: Stem superstars | Includes bibliographical references and index. | Audience: Grades K-1
Identifiers: LCCN 2019051754 (print) | LCCN 2019051755 (ebook) | ISBN 9781684508341 (hardcover) | ISBN 9781684046348 (paperback) | ISBN 9781684046386 (ebook)
Subjects: LCSH: Tyson, Neil deGrasse--Juvenile literature. | Astrophysicists--United States--Biography--Juvenile literature. | Astronautics--United States--Juvenile literature. | Universe--Juvenile literature.
Classification: LCC QB460.72.T97 C37 2021 (print) | LCC QB460.72.T97 (ebook) | DDC 520.92 [B]--dc23
LC record available at https://lccn.loc.gov/2019051754
LC ebook record available at https://lccn.loc.gov/2019051755

328N—072020
Manufactured in the United States of America in North Mankato, Minnesota.

★ Table of Contents ★

Chapter 1

Early Life

Neil deGrasse Tyson was born on October 5, 1958. He was the middle child of three. He grew up in New York City. He liked to stand on his family's roof. He wanted to study the stars.

Neil deGrasse Tyson grew up in the Bronx in New York City.

It was hard to see the stars. Tyson's parents took him to a **planetarium**. He took classes. His love for space began there.

Tyson loved seeing deep into space at the planetarium.

Tyson uses telescopes to study space.

Did You Know?
Tyson made money by walking dogs.

He saved for a **telescope**. Tyson set it up on roofs. Sometimes people called the police. They did not know why he was on the roof. But he used his telescope to show the police the stars.

Chapter 2

Space Studies

Tyson ran a science magazine for his high school. Then he went to Harvard University. He studied **physics** there.

Tyson graduated from Harvard.

Tyson went to the University of Texas next. People did not expect to see a black scientist. He faced **racism**.

Later in his life, other colleges gave Tyson special degrees.

Did You Know?

Tyson was in a Superman comic book.

Tyson became an **astrophysicist**. He studied space. There were only seven black astrophysicists in the country then.

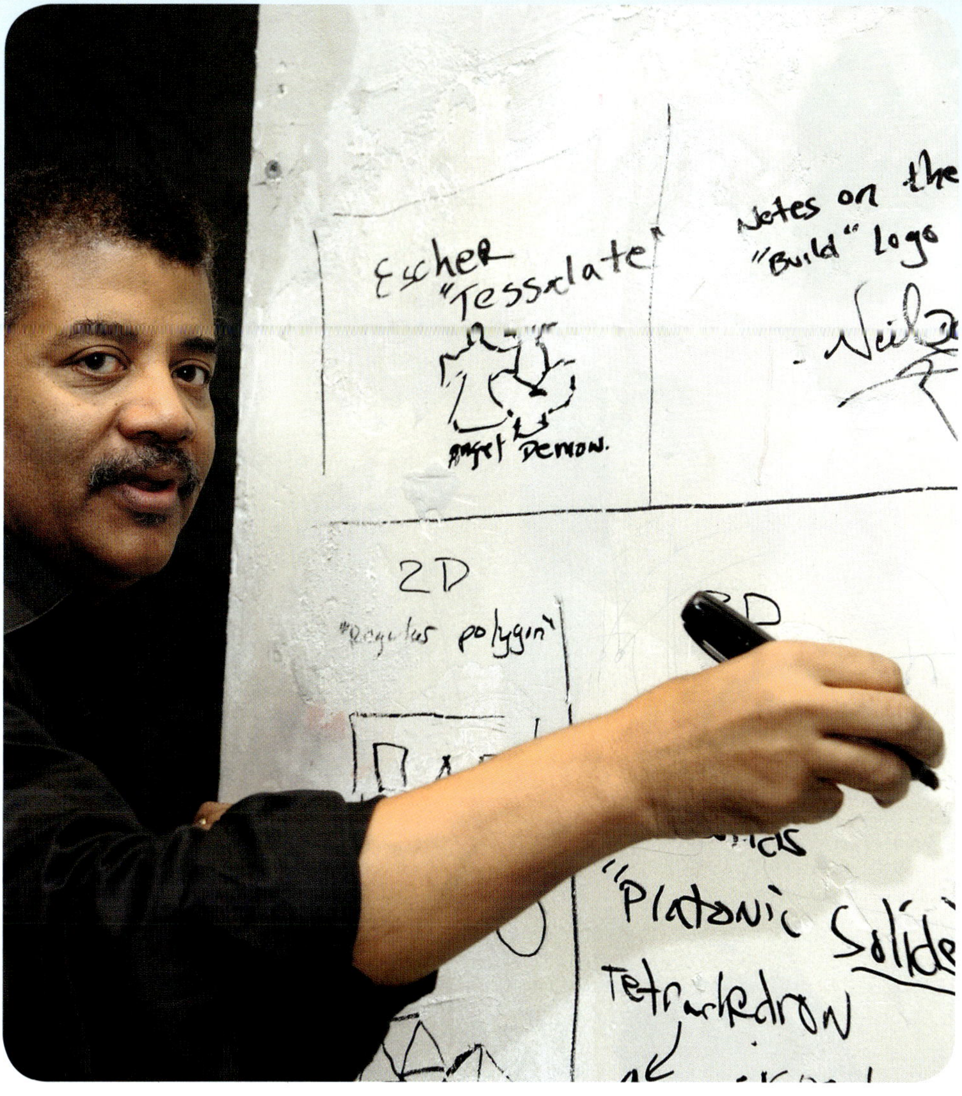

Tyson learns about space and teaches others too.

Chapter 3

Science Expert

Tyson studied groups of stars. Then he was hired to run the Hayden Planetarium. This is the same one he visited as a boy.

Tyson directs the Hayden Planetarium.

feedpop

Did You Know?

Tyson has an **asteroid** named after him.

Tyson is an author. He writes about space. He talks about science on TV. He also has a radio show.

Tyson has appeared on TV and radio.

Tyson likes teaching at the planetarium. And he likes teaching people everywhere about the stars.

Tyson meets with a group after giving a science talk.

★ Career Connections ★

1. Go stargazing! You can do this in your backyard or on a rooftop. Certain apps can help you identify constellations or locate a space station.

2. Ask an adult to take you to a planetarium or a science museum. These places often hold workshops or classes.

3. Some universities and science centers will let adults rent a telescope. See if you can try one out with the help of a parent or other adult.

4. Your local library may have science, technology, engineering, and mathematics (STEM) or **astronomy** books for children. Ask a librarian to help you find books about STEM and space.

5. Your school may have a science or math club. Consider joining it!

Glossary

asteroid (AS-tuh-roid): A rock that circles the sun.

astronomy (uh-STRON-uh-mee): The study of the stars, planets, and space.

astrophysicist (AS-troh-FIZ-uh-sist): A scientist who studies a branch of astronomy that deals with the laws of physics in space.

physics (FIZ-iks): The study of matter and energy and how they work together.

planetarium (plan-uh-TAIR-ee-uhm): A theater where you can see images of the night sky on the ceiling of a dome-shaped building.

racism (RAY-siz-um): When someone is treated unfairly because of race.

telescope (TEL-uh-skope): An instrument that makes distant objects seem larger and closer.

For More Information

Books

Norman D. Graubart. *Neil deGrasse Tyson*. New York: Enslow Publishing, 2016. This book teaches kids about Neil deGrasse Tyson's work.

Walt K. Moon. *Let's Explore the Stars*. Minneapolis, MN: Lerner Publications, 2018. This book teaches readers all about stars and outer space.

Websites

Astronomy: Our Place in Space

(https://www.amnh.org/explore/ology/astronomy) This page uses images, activities, and games to help kids learn about astronomy.

NASA Kids' Club

(https://www.nasa.gov/kidsclub/index.html) This page has images, games, and information about space for kids.

★ Index ★

★ About the Author ★

Rachel Castro is a Minneapolis-based writer. She holds degrees in English literature and creative nonfiction. When she's not writing children's books, she works for a public library and also teaches creative writing.